MathStart®
ANGLES

HAMSTER CHAMPS

by Stuart J. Murphy • illustrated by Pedro Martin

HarperCollinsPublishers

LEVEL
3

To Talia,
who is well on her way to becoming a champ
—S.J.M.

The publisher and author would like to thank teachers Patricia Chase, Phyllis Goldman, and
Patrick Hopfensperger for their help in making the math in MathStart just right for kids.

HarperCollins®, ♣ ®, and MathStart® are registered trademarks of HarperCollins Publishers.
For more information about the MathStart series, write to
HarperCollins Children's Books, 1350 Avenue of the Americas, New York, NY 10019,
or visit our website at www.mathstartbooks.com.

Bugs incorporated in the MathStart series design were painted by Jon Buller.

Hamster Champs
Text copyright © 2005 by Stuart J. Murphy
Illustrations copyright © 2005 by Pedro Martin
Manufactured in China by South China Printing Company Ltd.

Library of Congress Cataloging-in-Publication Data
Murphy, Stuart J., 1942–
    Hamster Champs / by Stuart J. Murphy ; illustrated by Pedro Martin.— 1st ed.
      p.   cm. — (MathStart)
    "Level 3: angles."
    ISBN 0-06-055772-9 — ISBN 0-06-055773-7 (pbk.)
    I. Angles (Geometry)—Measurement—Juvenile literature I. Martin, Pedro, 1967– ill. II. Title. III. Series.
QA482.M87 2005
516'.152—dc22
                                          2004022471
                                            CIP
                                            AC

Typography by Elynn Cohen   1 2 3 4 5 6 7 8 9 10   ❖   First Edition

Be sure to look for all of these **MathStart** books:

The front door slammed. A car drove away.

Pipsqueak poked her head out from under a pile of wood shavings.

"We're on our own," she said.

"Except for me," said Hector the cat.

"Stop that, Hector!" said Chuckles.

"Back off, fur face!" said Moe.

"Why are you always so mean?" asked Pipsqueak.

"There's nothing else to do when the people are gone," said Hector. "A cat has to entertain himself somehow."

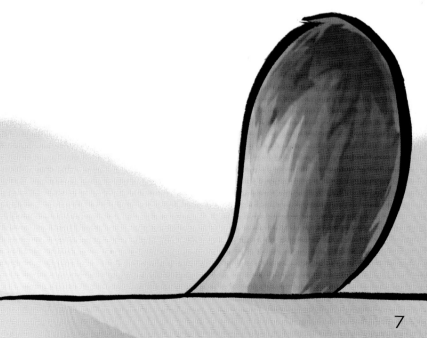

"What do you mean, nothing to do?" said Pipsqueak. "We're the hamster champs! And we've got a brand-new stunt!"

"Hah," snorted Hector. "What kind of a stunt can hamsters do?"

"Well, we could show you," Pipsqueak said. "But we'll have to get out of the cage."

"So you have to promise not to chase us," said Moe.

Hector thought for a minute.
"All right," he said. "Here's the deal.
If your hamster stunt is good enough,
I won't chase you. But if I get bored . . .
watch out!"

"No way you'll get bored," said Pipsqueak. "Hamsters, let's go!"

"Wait till you see what we can do, fleabag!" said Chuckles.

The hamsters lifted up the top of their cage and climbed out.

"I've found a car," yelled Moe.

"I found some blocks and a board that will work for a ramp," hollered Pipsqueak.

"Come and help me with this thing," called
Chuckles. "We'll need it to measure the angles."

"It's called a protractor," Hector grumbled.
"Don't you furry little snacks know anything?"

The hamsters piled up some pillows into a hill.

Pipsqueak placed the board so that one end was on the floor and the other was on top of some blocks. Then she carefully positioned the protractor so that the center was at the point of the angle.

"Thirty degrees," she announced.

The hamster team carried the car up the hill of pillows.
Then they started to run.

Pipsqueak jumped in first. The car went faster.

Moe was next. And then Chuckles landed right behind him. "Let's fly!" he hollered.

The car raced over the ramp and soared into the air.

Hector yawned. "You call that a stunt?" he asked.

"That was just a warm-up. Watch this!" said Pipsqueak.

Chuckles moved the ramp so that the angle was steeper. Again, Pipsqueak measured it with the protractor.

"The bigger the angle, the higher we'll fly," she said. "That's forty-five degrees."

The team zoomed down the hill, over the ramp, and into the air.

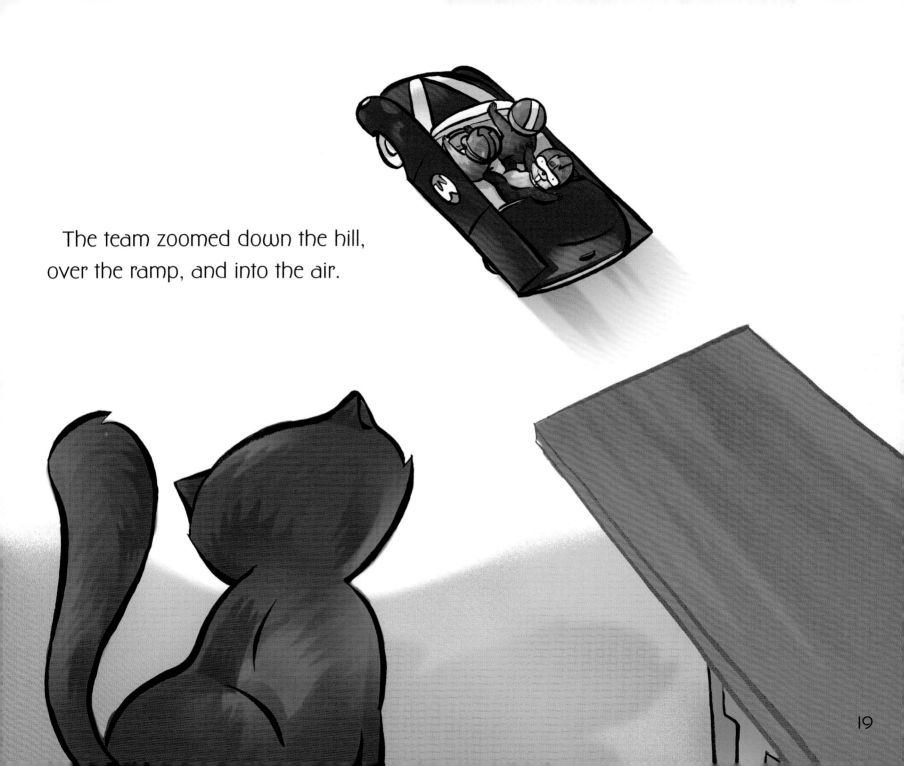

"I've seen better," said Hector. "Is that the biggest angle you can do?"
"Watch and wait!" said Moe.

20

Moe moved the ramp to make an angle of sixty degrees. He wiped the sweat from his whiskers. "That looks way too steep to me," he squeaked.

"We need the speed. We can do it!" said Chuckles.

The hamsters charged down the hill and started up the ramp. The car went slower . . . and slower . . . and then rolled backward all the way down the ramp.

"Oh, rats," said Moe.

"Hah!" said Hector. "Prepare to run!"

But Pipsqueak leaped to her feet, dusted herself off, and announced, "We missed that one, but we're going to try again—at an even larger angle!"

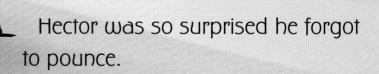

Hector was so surprised he forgot to pounce.

Chuckles and Moe stared at
Pipsqueak.

"Trust me," she said.

Chuckles held the the protractor.
Pipsqueak started to lift the ramp.
It went higher and higher, until
it was straight up—ninety
degrees.

"That's a right angle—like a wall," said Moe.
"We'll smash right into it!"

But Pipsqueak kept moving the ramp. Now the angle was larger than ninety degrees. Hector started to laugh.

But Pipsqueak kept going until the board was flat on the ground.

"One hundred and eighty degrees!" she yelled, and the hamsters ran to the top of the hill.

"A trick!" Hector snarled. "That's not a ramp. It's a straight line!"

"You wanted a bigger angle, and you got one," Pipsqueak yelled back.

Hector crouched. He lashed his tail. He jumped!
But the hamsters sped down the hill faster than he expected.
Hector plowed into the hill of pillows.
"Hamster Alert!" Pipsqueak yelled.

Chuckles climbed onto
Pipsqueak's shoulders.
Moe jumped up onto
Chuckles's.

As the car sped
past their cage,
Moe leaped
on top while
grabbing Chuckles's
hand.

Pipsqueak held on to Chuckles's tail.

The hamsters climbed back into their cage and pulled the top on tight.

And Chuckles said, "Ha, Hector! You'll never catch the hamster champs!"

31

In *Hamster Champs*, the math concept presented is measuring angles. An angle is formed when two line segments, or rays, meet an endpoint. The distance between the segments is measured in degrees. Learning about angles helps children identify and describe different geometric shapes.

If you would like to have more fun with the math concepts presented in *Hamster Champs*, here are a few suggestions:

endpoint — distance between is measured in degrees

- Read the story with the child and point out how to read the angles on the protractor as the hamsters make the ramp steeper.
- Reread the story and explain that angles are measured in degrees, and that a right angle has 90 degrees while a straight angle (two rays making a straight line) has 180 degrees.
- Use a clock to help the child understand the concept of an angle. Set a clock to 9:00. Ask the child what angle the two hands form. Have the child find other times when the two hands form a right angle. What angle is formed when it is 6:00? Now have the child find other times when the hands form a straight angle.
- The child and two friends can use a piece of string that is about six feet in length to make angles. Have one child hold one end, the second child the other end, and the third child hold the piece of string in the middle. Have them make 45°, 90° and 180° angles.

Following are some activities that will help you extend the concepts presented in *Hamster Champs* into a child's everyday life:

Body Angles: Have the child use her arms and legs to make angles of various sizes. What is the biggest angle she can make? Can she form a 90° angle by bending at her waist or knees? Can she stand up straight to form a 180° angle? What angles can she make with her wrists or ankles?

Playing a Robot: Have the child pretend he is a robot. As you give directions, he must follow them. For example, you might say: go forward five steps, turn 90° to the right, go backward six steps, turn 180° to the left, go forward three steps, turn 45° to the right, and so on.

Paper Airplane: Help the child fold a sheet of paper into a paper airplane. After each fold show the angles formed and have the child estimate and then measure the angles.

The following books include some of the same concepts that are presented in *Hamster Champs*:

- SHAPE UP! by David A. Adler

  - PIGS ON THE BALL: *Fun with Math and Sports* by Amy Axelrod

  - SIR CUMFERENCE AND THE GREAT KNIGHT OF ANGLELAND: *A Math Adventure* by Cindy Neuschwander

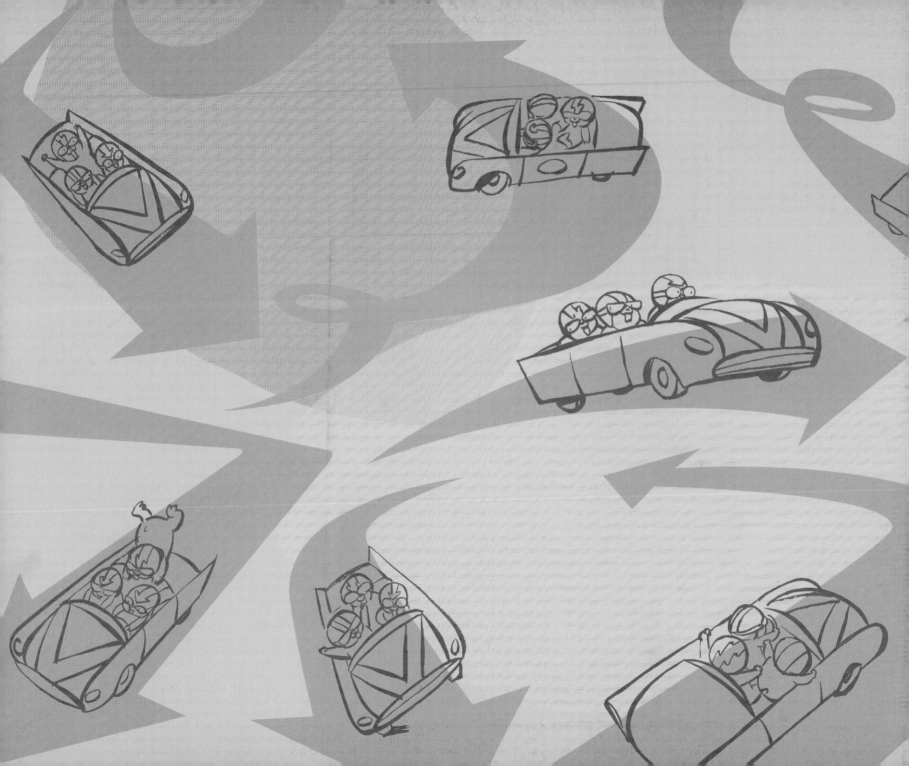